LEARN
STEM
FROM
BASKETBALL

LEARNING
STEM
FROM
BASKETBALL

Why Does a Basketball Bounce?
And Other Amazing Answers for Kids!

MARNE VENTURA

Sky Pony Press
New York

Sky Pony Press books may be purchased in bulk at special discounts for sales
promotion, corporate gifts, fund-raising, or educational purposes. Special
editions can also be created to specifications. For details, contact the Special
Sales Department, Sky Pony Press, 307 West 36th Street, 11th Floor, New York, NY
10018 or info@skyhorsepublishing.com.

Sky Pony® is a registered trademark of Skyhorse Publishing, Inc.®, a Delaware
corporation.

Visit our website at www.skyponypress.com.

10 9 8 7 6 5 4 3 2 1

Manufactured in China, February 2021

This product conforms to CPSIA 2008

Library of Congress Cataloging-in-Publication Data is available on file.

Cover design by Mona Lin
Cover photographs: Getty Images

Print ISBN: 978-1-5107-5701-1
Ebook ISBN: 978-1-5107-6048-6

Printed in China

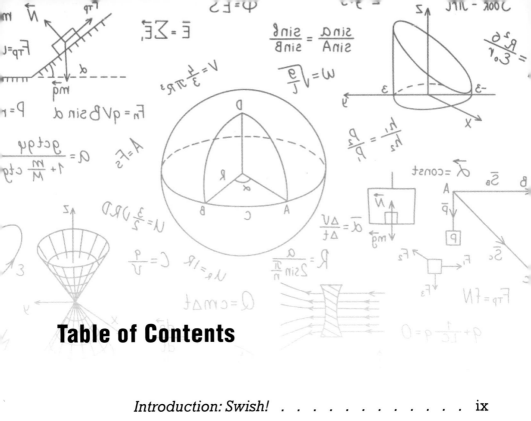

Table of Contents

Introduction: Swish! ix

Chapter One: The Spectacular Science of the Sport

Why does a basketball bounce? 1
Why do players jump when they shoot? 3
What is hang time? 5
What's muscle memory? 7
How does gravity affect a pass? 9

Chapter Two: Tremendous Technology Tidbits

When were electronic scoreboards invented? . . . 11
What's a smart basketball? 12
How does technology make basketball more
        fun for fans? 15
Why are there cameras in the rafters of
        NBA arenas? 17
How do referees use technology? 19

Chapter Three: Engineering Awesome Equipment

What are basketballs made of?. 21

What's a breakaway rim? 23

Why does a basketball rim have a net? 25

Why do basketball courts have wood floors?. . . . 27

How have basketball shoes changed over
      the years? 29

Chapter Four: Math Mysteries Solved

What's the best angle for a shot? 31

Are all basketball courts the same size? 33

What statistics do basketball players use? 34

Why is a shot clock 24 seconds long? 37

How much money do pro basketball players
      earn? 39

Glossary 41

Books and Websites for Further Reading . . . 43

About the Author 44

Endnotes 45

Introduction: Swish!

City parks, country schoolyards, college gymnasiums, neighborhood driveways—it's hard to travel very far without seeing a basketball hoop. People everywhere shoot hoops for fun, for exercise, with a team, or to spend time with friends. Basketball is the only major sport that was invented completely in the United States of America.[1]

James Naismith was a physical education teacher at the Young Men's Christian Association (YMCA) Training School (now Springfield College) in Massachusetts. In December 1891, Naismith was in charge of a class of 18 restless students who enjoyed outdoor sports such as football and lacrosse. Naismith's job was to come up with an indoor game that would keep the students busy and entertained during the winter months. When he nailed two peach baskets to opposite ends of the gym balcony, divided his students into two teams and gave them a soccer ball, the game of basketball was born.[2]

A lively game of basketball is science, technology, engineering and math (STEM) in action. STEM concepts can be found in the physics behind the bouncing of the ball, the workings of the electronic scoreboard, the engineering of a better sneaker, and the math behind a shot clock.

Chapter One

The Spectacular Science of the Sport

Why does a basketball bounce?

Imagine trying to dribble a basketball that hasn't been inflated. Not much bounce, right? Basketballs bounce because they are filled with pressurized air and made with elastic materials.

Air is made up of constantly moving **molecules**. Although we can't see them, they are busy vibrating, bumping into each other and everything around them. Air pressure is the weight

of these molecules pressing against their surroundings. More molecules in a given space create higher air pressure. Less molecules in the same space create lower air pressure.

When a basketball hits the court floor, it pushes on the wood and the wood pushes back. The bottom of the ball flattens slightly. The air molecules inside the ball are squeezed into a smaller area. The crowded molecules respond by pushing back out. As they move apart, the ball returns to its round shape. Like a spring, this action pushes the ball back upward in a bounce.[1]

Air pressure is measured by pounds per square inch (PSI). National Basketball Association (NBA) rules call for basketballs to be inflated between 7.5 and 8.5 PSI.[2] This amount of air pressure gives the ball just the right amount of bounce.

Temperature

Air molecules expand and speed up as they warm. Their energy increases. When cold, they contract and slow down. This is why a warm ball will bounce higher than a cold one.[3]

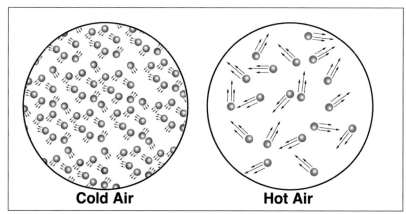

Cold Air   Hot Air

Credit: Fred Ventura

Why do players jump when they shoot?

Stephen Curry is one of the best shooters in the NBA.[4] But how does this point guard for the Golden State Warriors do it?[5]

It takes both vertical and horizontal **force** to get the ball up and over to the basket. When a player bends his knees and pushes his feet against the court, the court pushes back and the player rises in a jump. The **energy** from his rising body is partly transferred to the ball. This helps the ball move higher and faster toward the basket. It also allows the player to release the ball closer to the basket.[6] Another reason for the jump is to get the ball above the player guarding the shooter, which enables them to better see the basket.

Have you noticed that Stephen Curry does not jump straight up and come straight down when he shoots? After Curry releases the ball, he comes down to the ground with the upper part of his body angled slightly back, and his lower body slightly forward. By relaxing his body in this way, he transfers more energy from his body to the ball. This gives the ball more force to reach the basket.[7]

Kenny Sailors

Kenny Sailors introduced the jump shot in the 1930s. He formed the habit of jumping because he played with his older, taller brother. The only way for him to make a shot was to jump.[8]

What is hang time?

Some pro basketball players seem to hang in midair when they jump for a slam dunk. Can LeBron James and Michael Jordan defy **gravity**? Do they have a super power? Super, yes! Supernatural, not at all. The illusion of hang time can be explained by the laws of physics.

Hang time is the total time that a player stays in the air during a jump. It starts when his feet leave the ground, and ends when his feet touch back down. How high a player jumps depends on the amount of force created by his leg muscles pushing against the court. The more force, the higher the jump. The higher the jump, the longer the hang time.[9]

As the player rises, his speed slows. When he reaches the top of the jump, gravity pulls him back down. His speed increases as he drops to the ground. His speed is faster at the beginning and end of the jump, and slower during the middle of the jump. For this reason, about 70 percent of the jump time is spent in the top half of the jump. That's why it looks like he's hanging in the air.

Top Half of Jump
70% of Time

Bottom Half of Jump
30% of Time

Credit: Fred Ventura

How Long?

An average person's hang time is about 0.53 seconds. A vertical jump of four feet leads to hang time of about one second. Michael Jordan's longest hang time record is 0.92 seconds.[10, 11]

What's muscle memory?

Muscle memory allows you to do certain actions without really focusing on each step. The expression "It's like riding a bicycle" describes muscle memory. Once you learn to ride a bike, you don't think too hard about it—you just hop on and pedal. Forming a habit is another name for muscle memory. Scientists have shown that there are actually changes in the shape of your brain that occur during muscle memory formation.[12, 13]

Skilled basketball players practice shooting and dribbling over and over again. They want to lock in the right form and action so that, during a fast game, their muscle memory will kick in and help them score. Before a game, players warm up to activate their muscle memory. They take shot after shot from the free throw line. They move around the floor and shoot from other spots as well.

The memory is not really in the muscles. It's in the link between the brain and the muscles—the nervous system. The more often a player successfully shoots and dribbles, the stronger the pathways between his muscles and brain become. The better the muscle memory, the more a player can free up his attention to focus on other parts of the game.[14] [15]

Practice Makes Perfect

The old saying "practice makes perfect" refers to muscle memory. Because of muscle memory, it's important to practice using the correct form. Muscle memory can work against you if you're repeating mistakes.[16]

How does gravity affect a pass?

Gravity is the force that pulls objects toward the earth. Gravity gives objects weight. Every move a basketball player makes is affected by gravity. Passing, dribbling, shooting, dunking—during all of these actions, the force of gravity is pulling the ball toward the court.[17]

When a player passes the ball, he applies force to the ball with his hands and arms. This force pushes the ball toward the player who will receive the pass. Aside from a bounce pass, a good player will automatically send the ball in an arc, aiming

slightly higher than where he wants it to land. This is because he is taking the downward pull of gravity into account. If he did not put a bit of upward force on the ball, it would land lower than he intended.[18]

The force of gravity on an object depends on its **mass**. The greater the mass of an object, the greater the force of gravity. A basketball and a bowling ball are similar in size, but the bowling ball has more mass. Imagine passing a bowling ball to a teammate. You would need to apply more force, and more upward lift, to compensate for the larger pull of gravity.[19, 20]

Mass versus Weight

Weight is a measure of the pull of gravity on an object. Mass is a measure of how much matter an object has.[21] To measure mass, scientists use a balance. To measure weight, scientists use a scale.[22]

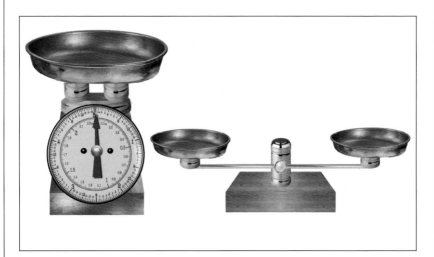

Credit: Fred Ventura

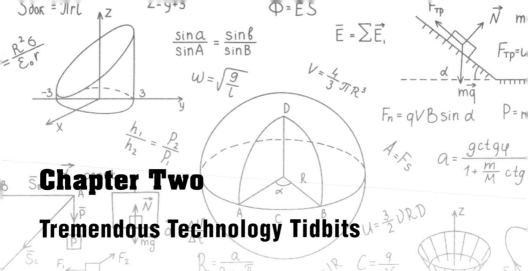

Chapter Two

Tremendous Technology Tidbits

When were electronic scoreboards invented?

Fans at basketball games watch the scoreboard as much as they watch the players. The scoreboard shows how many points each team has earned. It tells which of four quarters (two halves in cases such as college basketball and WNBA) the game is in, and how much time is left in that quarter. It tells how many fouls, and the number of timeouts remaining for each team.[1]

In the past, scoreboards were equipped with a **mechanical** clock and cards with numbers on them to show the score. When a player earned points, the person who was the scorekeeper would find the card with the correct number and put it on a hook on the scoreboard.[2,3] In the 1950s, several wires connected **electronic** scoreboards to a table where the scorekeeper sat. The scorekeeper changed the display on the scoreboard manually from the table.[4, 5]

Modern, wireless scoreboards are now controlled with computers.[6] Big screens hang from the ceilings of basketball arenas. An operator changes the score on the screen from a control panel.[7] Scoreboards display players' pictures, names, and **statistics**. They also have a horn or buzzer to automatically signal things like the end of a quarter (or half) or a foul.[8]

LED

In 1962, scientist Nick Holonyak Jr. created the light-emitting diode (LED). LED technology makes digital scoreboards possible. LEDs are tiny light bulbs that fit into an electrical circuit. On a digital scoreboard, LEDs form the numbers that show the score.[9]

What's a smart basketball?

Today we use smartphones to send and receive emails and messages and check the weather. We ask our smartphones for directions and information. Basketballs can be smart, too! Technology companies are making basketballs that use electronic **sensors** to give players feedback on their shooting and dribbling. These measurements help players practice the moves that work best.[10]

One smart basketball contains a sensor that connects wirelessly to a smartphone app. It tracks how many times a player tries to score, and how many times he succeeds. It records how long it takes to make the shot. It even makes sounds, like the crowd cheering or the clock buzzing, so players can practice as if they are in a live game.[11]

Credit: Fred Ventura

Another smart basketball helps players practice dribbling. It contains an **optical** meter that tracks the ball's movement. It gives players feedback on how long they keep up the strength of their dribble, how fast they change direction, and whether they are slowing down too much. Players get feedback on a smartphone app. The app gives live, **real-time** audio feedback, as if a coach were giving advice and guidance.[12]

Computer Vision
Basketballs that give feedback to a player use computer vision. The player places his smartphone on a stand on the floor. As the player dribbles, a computer program works with the phone's camera to "watch" the ball by finding its microfiber surface.[13]

How does technology make basketball more fun for fans?

Technology is the use of science in everyday life. Virtual reality (VR) is a world created by a computer program, such as a video game. Augmented reality (AR) occurs when a computer program adds virtual reality to the real world. Experts are finding new ways to make basketball more fun for fans with the use of these technologies.[14]

One smartphone app lets fans feel like they are walking through a 3D video of a recorded basketball game. The Golden State Warriors place VR cameras in a front-row seat during the game. Fans can use the video to experience being in the middle of the action.[15]

Currently, six NBA teams use 360-degree video replays in their arenas to help fans see what is happening on the court. The video is delivered to the fans' smartphones. Fans can review shots and plays. They can change the angle in order to get a better look at the action. Cameras also send video to the scoreboards above the court so fans can watch replays on the big screen.[16]

Smart Stadiums

Another cool use of technology for basketball fans is the "smart stadium." Fans can use their smartphones to get directions, updates on player statistics during the game, and other useful information.[17] Fans can even order food and drink from their seats using smartphones.[18]

Why are there cameras in the rafters of NBA arenas?

Every arena in the NBA has now installed a high-tech camera tracking system above the court. This system delivers **data** to coaches, players, and fans. Teams use these statistics to play better and win more games.[19]

In the past, for example, coaches told their shooters to get close to the basket to shoot. They thought shooting from behind the three-point line was too risky. But after studying statistics from tracking systems, coaches changed their strategy. They learned that shooting more often from behind the three-point line leads to more wins.[20]

How does this camera tracking system work? Six cameras are mounted evenly above the court. They record all the player and ball movements. They tell who scored each basket, what kinds of shot was made or missed, and each players' position and speed. A sensor on the ball and tiny sensors worn by the players connect wirelessly to the cameras. Data from the system gives coaches and players feedback they can use to play better. Teams use the data to compare player performance, monitor their health after an injury, and plan their game strategy.[21]

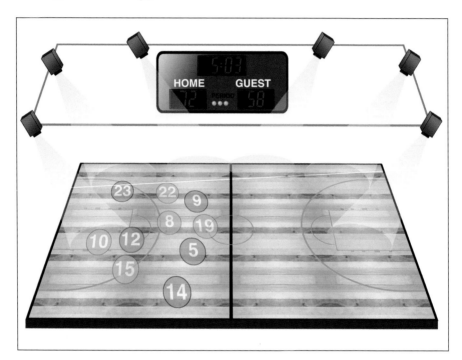

Credit: Fred Ventura

High Speed Data Collection
NBA arena cameras capture the 10 players and the ball 25 times per second! During one game, the system records more than 72,000 data points on the court![22]

How do referees use technology?

Being an NBA referee is a difficult job. For example, if a player takes a shot and the end-of-quarter (or half) timer buzzes, the ref has to decide if the ball left the player's hands before or after the buzzer sounded. At the end of a close game, this decision can decide which team wins or loses.[23]

Technology has made refs' jobs easier. Instant video replays from different angles show whether a shot was released in time. Video recordings also show if the shooter's feet were behind the three-point line. Red LED lights in the backboards help refs, too, as they go on when the game clock expires. Lights along the sidelines of the court floor also help refs see where the ball is when the timer buzzes.[24]

In the past, a person acted as sideline timekeeper. They listened for the ref to blow his whistle, and then they shut off the clock. Today, NBA refs use a system with a microphone near their whistle and a belt pack on their waist. When the ref blows the whistle, the system sends a radio signal that stops the clock. The ref presses a little button on their belt to restart the clock.[25, 26]

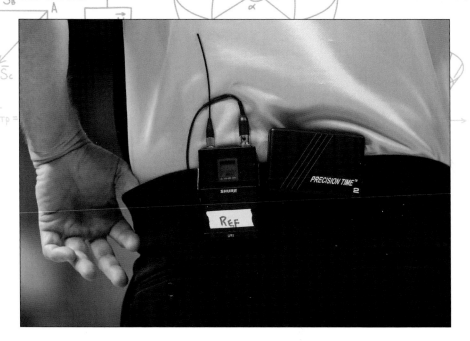

Speed of Light
Radio waves that travel from the ref's whistle to the scorer's table to stop the clock move at the speed of light. A human timekeeper takes about 0.7 seconds to react and stop the clock by hand.[27]

Chapter Three

Engineering Awesome Equipment

What are basketballs made of?

The first basketball players used smooth soccer balls. In 1894,
A. G. Spalding made the first official basketball.[1] The outside
of the ball is leather. The inside is string-wrapped rubber.[2]

Official NBA basketballs are made according to strict rules. Rubber is melted and formed into long, flat sheets. These sheets are then folded into quarters and cut into circles. The edges are bonded together, which forms the inner bladder. The bladder is heated and filled with air. Next, it's wrapped with nylon string, similar to a golf ball. This makes the bladder rounder and stronger. Finally, eight leather panels are glued together by hand to cover the ball.[3]

The balls are inflated and measured. They must be between 29⅝ and 29¾ inches (75½ and 75.57 cm) in circumference. Next the balls are rebounded at 20 miles per hour (32 km/h) through a machine that measures how well they bounce back.[4]

Game inventor James Naismith and sporting goods maker A. G. Spalding came up with the idea of putting bumps, called *pebbling*, on the basketball. The bumps create friction between the ball and the player's hand. This makes it less slippery, so it's easier to handle.[5]

Why Orange?

Basketballs were originally dark brown. Tony Hinkle was a college basketball coach at Butler University in Indiana in the 1950s. He noticed that brown balls were hard to see during a game. He worked with the Spalding Company to make orange basketballs that are more visible.[6]

What's a breakaway rim?

Baskets have come a long way since the invention of the game in 1891. The peach basket was replaced by a metal rim with a net in 1901. In 1904, players started using wooden backboards behind the hoop. The first glass backboards were introduced in 1909. Transparent glass let the audience see more of the game.

Glass backboards often shattered when players grabbed the metal rim while jumping up to dunk. Players would injure their wrists, and the game would have to be stopped while the broken glass was cleaned up.

In 1975, Arthur Ehrat worked at a grain elevator in Lowder, Illinois. His nephew was an assistant coach for the Saint Louis University basketball team. Knowing his uncle was handy with machines, Ehrat's nephew asked him to design a rim that could bend and spring back when 125 pounds (57 kg) of force pulled on it. To make the breakaway rim, Ehrat used bolts, metal braces, and the spring from a John Deere digging tractor. Because the rim is flexible, it doesn't break when a player tugs on it. This means players can entertain fans with slam dunks without breaking the backboard or hurting themselves. [7, 8]

Before Nets

In the first days of basketball, when a player made a basket, someone had to climb a ladder to retrieve the ball. Another method was to poke a broomstick through the woven peach basket to push the ball out the top.[9]

Why does a basketball rim have a net?

Why are nets hung from hoop rims? Why not just shoot the ball through the metal rim? It wasn't until 1901, 10 years after the invention of basketball, that metal rims were used instead of wooden peach baskets. The rim had a net with a closed bottom. In 1906, someone had the good idea of cutting out the bottom of the net so the ball could fall through on its own.[10, 11]

The contrast between the white net and the orange ball makes it much easier for the referee to see whether the ball went through the hoop or not. The net also slows the ball down a bit as it falls, and this also helps the ref see that the player truly scored.

The net also causes the ball to fall downward rather than flying off in a sideways direction. This makes it easier to retrieve, and keeps the game moving.[12]

Without a net there could be no *SWISH!* That's the awesome sound the ball makes as it catches for a split second in the narrow part of the net and then drops through.

Rims and Nets

Basketball rims are made of ⅝ inch (1.6 cm) steel diameter rod. Nylon nets are hung from the bottom of the rim with lengths between 15 and 18 inches (38.1 and 45.7 cm). Rims are mounted 10 feet (3.05 m) from the court floor.[13]

Why do basketball courts have wood floors?

The gym floor at the first game of basketball in 1891 was made of maple wood planks. Today, 29 out of 30 NBA arenas have maple floors. The Boston Celtics' arena floor is made of red oak. The NBA favors maple because of its hardness, strength, springiness, and cost.

Timber companies rate woods on a scale from zero to 4,000. Hard maple has a rating of 1,450. The main reason for it's use is due to the fact that it is strong and resists scratches. At the same time, it's soft enough to sink slightly and spring back. Because it's slightly flexible, it's less likely that players will injure their feet, ankles, or legs.

Another advantage to hard maple is the tightness of the wood grain. Because the fibers are close together, the wood has less cracks to trap dirt and dust. This makes it easier to keep clean, and so it stays smoother.[14]

Hard maple wood is very light in color. This helps provide a sharp contrast between the floor and the dark orange-brown ball. Contrast makes it easier for players, refs, and fans to see the ball as it moves. Also, the light color reflects the lights in the rafters, helping brighten the arena.[15]

Air Discs

In the same way that sneaker makers put small discs filled with air in the soles of shoes, engineers lay small, rubber, air-filled discs under wooden basketball courts. This makes the court springy, and helps prevent injury and fatigue.[16]

How have basketball shoes changed over the years?

The first basketball teams ran around the court in their everyday leather shoes. It wasn't until 1917, 26 years later, that the Converse company made shoes just for basketball. They called them Chuck Taylor All Stars after the popular player.

The shoes, still made today, have rubber soles and canvas tops that cover the ankle. Rubber works much better than leather on a wood court, as it grips the wood and keeps the player from sliding. The waffle-like grid in the sole helps with grip as well. The high-top design supports the players' ankles as they run, dribble, and land after a jump shot.

The tradition of naming shoes after basketball

stars continued. In 1973, the Puma company named a shoe the "Puma Clyde" after NBA star Walt "Clyde" Frazier. This shoe had a suede upper and a wide sole for stability. Around the same time, Nike came out with leather high tops called "Air Jordans" for NBA star Michael Jordan. In 1991, Reebok made a "Pump" shoe that players could inflate for extra ankle support. Today, basketball shoes are very colorful. Engineers have found ways to make them lighter and more supportive.[17]

Not Just for Basketball

Since the first All Stars, basketball shoes have had a big influence on footwear for everyone. Fans want to wear the same shoes that their favorite stars wear, and are willing to pay high prices for them.[18]

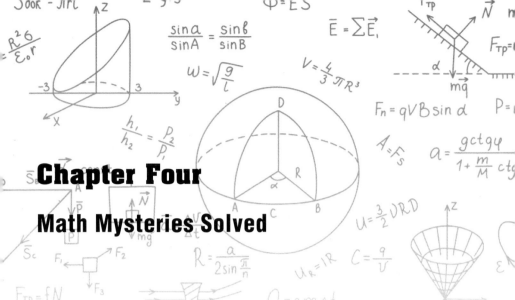

Chapter Four

Math Mysteries Solved

What's the best angle for a shot?

A ball thrown straight down into the basket from above—a 90-degree angle—has plenty of room to go through. But players can't always dunk a ball straight down for a **swish**. They usually throw the ball so that it travels in an arc. Experts who have studied players in action say that if the ball is shot

Credit: Fred Ventura

straight, and aimed about 11 inches (28 cm) beyond the front of the rim, that an arc where the ball enters the basket at a 45-degree angle has the best chance of scoring.[1]

It might seem like a higher arc is a good strategy, but players must use more force to throw the ball in an arc where the entry angle is higher. This makes it more likely that they will lose control of the ball and throw it short or long. The force with which a high arc is thrown also makes the shot less soft. This is a problem because a soft shot is more likely to bounce gently off the rim and go through the net. A more forceful throw is more apt to bounce out of the rim.[2]

Rim Area

Imagine holding the rim perpendicular to the floor and tilting it until it's parallel to the floor. See how the visible area inside the rim decreases? In the same way, the angle of entry changes the area that the ball can go through.

Are all basketball courts the same size?

Professional basketball courts, as well as college, high school, and junior high courts, can be different in their overall size and layout. The actual court markings for the foul line and the backboard and rim are always the same. For NBA courts, the overall court is 94 feet long and 50 feet wide (28.7 x 15.2 m). The free throw line is 15 feet (4.6 m) away from the front of the backboard. The backboard extends out four feet over the baseline into the key. A six-foot (1.8 m) arc extends from the foul line away from the basket to form the key.

The rim is 10 feet above the court floor. Backboards are six feet wide by three and one-half feet tall (1.8 x 1 m). All rims are 18 inches (45.7 cm) in diameter. The inner square on the backboard is 24 inches wide by 18 inches tall (61 x 45.7 cm). All the lines marked on the floor are two inches (5 cm) wide.

The position of the three-point line or arc differs by organization. The measurement is taken from the center of the rim to the top of the arc.

NBA: 22 feet (6.7 m) from the center of the rim to the corners of the arc, and 23 feet 9 inches (7.3 m) elsewhere.

WNBA and International: 20 feet 6 inches (6.3 m) to the top of the arc.

NCAA Men's: 20 feet 9 inches (6.4 m) to the top of the arc.

The restricted arc is the semicircle drawn under the rim. If a defensive player's feet go inside this arc, they cannot draw a charging foul. This arc has a radius of four feet (1.2 m) from the center of the basket.[3, 4]

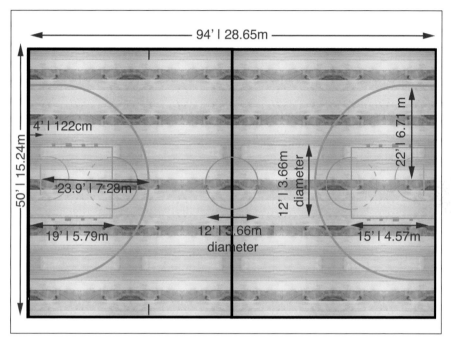

Credit: Fred Ventura

Scoring
Any shot made from outside the arc scores three points. Shots made from inside the arc score two points. A free throw shot is worth one point.[5]

What statistics do basketball players use?
Data are facts and numbers that researchers collect. Statistics is another word for data. In basketball, analysts study the numbers they collect using video systems in the rafters of the

arenas. They use math and computer programs to compare how different players perform. They identify actions that lead to more wins. They make predictions about which teams or players will score.[6]

One important statistic is Field Goal Attempts (FGA). FGA measures how many times a player takes a shot. Analysts who study basketball data say that the higher a team's FGA, the higher the chance of scoring and winning.

Effective Field Goal Percentage (EFG%) is another important statistic. This number accounts for the fact that a three-point field goal is worth more than a two-point field goal. The formula is eFG% = (FGM + (0.5 * 3P))/FGA. This stat shows how well a player shoots and scores. Below is an example showing LeBron James's EFG% for the Cleveland Cavaliers in the 2017–18 season:

Field Goals Made (FGM): 857
Two-point Field Goals (2P FG): 708
Three-point Field Goals (3P FG): 149
Field Goals Attempted (FGA): 1580

857 + (0.5 * 149) / 1580
(857 + 74.5) / 1580
931.5 / 1580
0.5895569

The eFG% for LeBron James in his 2017–18 season was 0.590.

Free Throw Attempts (FTA) tells how often a player gets to the free throw line. Free Throw Percentage (FT%) is found by dividing the number of successful free throws by the number of FTAs.[7]

By counting, recording, and studying these numbers, coaches can compare players' performance. Players can identify areas where they need to improve.

Which shot to take?

By crunching the numbers, analysts have found that it's often better to take a risk and try for a three-point shot than to play it safer and go in for a two-point shot.

Why is a shot clock 24 seconds long?

Before the NBA began using a shot clock, basketball was a much different game. Teams would get the lead, and then just stall by dribbling or passing the ball until the time ran out. Games could be pretty boring, and many ended with neither team having very many points. For example, in 1950, the Fort Wayne Pistons beat the Minneapolis Lakers, 19–18.

The owner of the Syracuse Nationals, Danny Biasone, had an idea. Why not set a time limit on how long a team could keep the ball without trying to shoot? He ran it by the NBA and they decided to give it a try. In 1954, the first NBA game to use the shot clock ended when the Rochester Royals beat the Boston Celtics, 98–95.[8]

Why did Biasone decide to make the time limit 24 seconds? He studied the action-packed games that he found most fun to watch. He noted that each team took about 60 shots. So, for two teams, he allowed 120 shots during the game. A game is four 12-minute quarters, or 48 minutes. 48 minutes is 2,880 seconds. 2,880 seconds divided by 120 shots is 24.[9]

Virtual Shot Clock

In 2019, Turner Sports began using on-court virtual shot clocks. Once a team gets the ball, a picture of a shot clock is projected onto the court floor. The virtual clock helps fans watch the players as well as the clock.[10]

Credit: Fred Ventura

How much money do pro basketball players earn?

The average NBA player salary for the 2019–20 season was $7.7 million. The season before, the average salary was nearly $6.4 million. Some players are paid more because they perform so well and are considered stars. For example, Stephen Curry earned $37.5 million in the 2018–19 season, and $40.2 million in the 2019–20 season.[11, 12]

The average American worker earned a bit less than $50,000 in 2019.[13] Why does the top NBA player make 800 times more than the average American worker?

Basketball is a multi-billion-dollar industry.[14] Fans are willing to pay a lot of money to watch NBA games. The average price of a ticket to a Los Angeles Lakers game was $549 in 2019. The average NBA arena has about 20,000 seats. In addition to earning money by ticket sales, NBA owners earn money by selling basketball merchandise and food. Television networks pay the NBA in order to broadcast the

games. Advertisers pay them to place ads in the arena. When a player is popular, more fans pay to come and watch him play.[15]

Shoe Deals

Superstar basketball players sometimes earn extra money by endorsing or promoting a certain brand of shoe. For example, Michael Jordan signed a deal with Nike to make Air Jordan sneakers.[16] Shoe manufacturers pay players anywhere from thousands to hundreds of millions of dollars.[17]

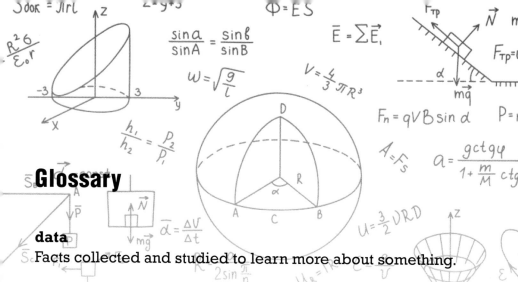

Glossary

data
Facts collected and studied to learn more about something.

electronic
Using components that control an electric current to work.

energy
The power or ability to move, or to be active.

force
A strength, or energy, or power.

gravity
The pull of the mass of something, such as Earth.

mass
The amount of matter in an object.

mechanical
Doing work by a combination of moving parts.

molecule
The smallest particle of a substance that has the properties of that substance.

optical
Related to the eye or to vision.

real time

The actual time during which something happens.

sensor

A device that detects or measures something.

statistics

The branch of mathematics that deals with collecting and studying number data.

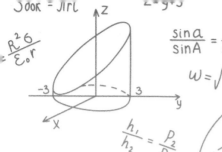

Books and Websites for Further Reading

Pernu, Dennis. *In Focus Sports: Basketball.* Minneapolis, MN: Lerner Books, 2019.

Smibert, Angie. *STEM in Basketball.* North Mankato, MN: ABDO, 2018.

Williams, Heather. *Basketball: A Guide for Players and Fans.* Mankato, MN: Capstone Publishing, 2020.

Basketball STEM Project:
https://basketballstem.weebly.com/index.html

Chevron Stem Zone: Basketball:
https://www.chevron.com/stemzone

How Basketball Works:
https://entertainment.howstuffworks.com/basketball.htm

About the Author

Marne Ventura is the author of more than one hundred books for children, including the companion to this book, *Learning STEM from Baseball*. A former elementary school teacher, she holds a master's degree in Reading and Language Development from the University of California. Marne's nonfiction titles cover a wide range of topics, including STEM, arts and crafts, food and cooking, biographies, health, and survival. Her fiction series, the *Worry Warriors*, tells the story of four brave kids who learn to conquer their fears. Marne and her husband live on the central coast of California.

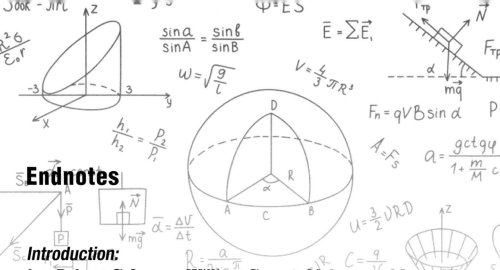

Endnotes

Introduction:

1. Robert G. Logan, William George Mokray, and Larry W. Donald, "Basketball," *Encyclopedia Britannica,* Mar. 25, 2020, https://www.britannica.com/sports/basketball.
2. "The Birthplace of Basketball," *Springfield College,* https://springfield.edu/where-basketball-was-invented -the-birthplace-of-basketball.

Chapter 1:

1. "Do Basketballs That Are Fully Inflated Bounce Better Than Flatter Ones?" *UCSB ScienceLine,* Apr. 17, 2013, http://science line.ucsb.edu/getkey.php?key=3939.
2. "What Is the PSI of a Regulation Basketball?" *Reference,* https://www.reference.com/world-view/psi-regulation -nba-basketball-20a203be4fcb0ae2.
3. "Q and A: Hot and Cold Basketballs." *Department of Physics University of Illinois at Urbana-Champaign,* Oct. 22, 2007, https://van.physics.illinois.edu/qa/listing.php ?id=107&t=hot-and-cold-basketballs.
4. Kirk Goldsberry, "The Absolute Best Shooters of This NBA Decade," *ESPN,* Sep. 16, 2019, https://www.espn.com /nba/story/_/id/27587041/the-absolute-best-shooters -nba-decade.
5. Chris Ballard, "Golden Touch: How to Shoot Like Steph," *Sports Illustrated,* Mar. 14, 2016, https://www

.si.com/nba/2016/03/14/stephen-curry-golden
-state-warriors-bruce-fraser-steve-kerr-shot.

6. Zhijun Sun, "The Science Behind Shooting a Basketball,"
    Science in Our World: Certainty and Controversy." Sep.
    13, 2015, https://sites.psu.edu/siowfa15/2015/09/13/the
    -science-behind-shooting-a-basketball/.

7. "The Physics of a Jumpshot," *Scot Shot Basketball*, Apr. 7,
    2017,https://www.youtube.com/watch?v=M1j30ebUvZE.

8. Karen Given, "Kenny Sailors: The Long-Forgotten Inventor
    of Basketball's Jump Shot," WBUR (Boston's NPR), Feb. 20,
    2016, https://www.wbur.org/onlyagame/2016/02/20/
    kenny-sailors-jump-shot-basketball.

9. Wendy Stewart, "The Physics of Basketball Player
    Hang Times," *SportsRec*, Nov. 16, 2018, https://www
    .sportsrec.com/7300810/the-physics-of-basketball
    -player-hang-times.

10. Ashish, "The Science of Michael Jordan's Slam Dunks
    and Hang Time," *Science ABC*, Jul. 15, 2015, https://www
    .scienceabc.com/pure-sciences/secret-michael-jordan
    -slam-dunks-basketball-math-physics-hang-time-jump
    .html.

11. "Sport! Science," *Exploratorium*, https://www.explorato
    rium.edu/sports/faq11.html.

12. Coach Al Heystek, "Free Throw Coaching Tip on Activating
    and Developing Muscle Memory to Improve Free Throw
    Shooting," *Free Throw Trainer*, http://freethrowtrainer
    .com/train-muscle-memory-for-improving-free-throw
    -shooting.html.

13. "The Amazing Phenomenon of Muscle Memory," *Oxford
    University*, Dec. 14, 2017, https://medium.com/oxford
    -university/the-amazing-phenomenon-of-muscle-memory
    -fb1cc4c4726.

14. "Muscle Memory," *Advantage Basketball Camps*, https://www.advantagebasketball.com/muscle_memory.htm.

15. "The Science of Muscle Memory and How It Works in Your Favor," *Cathe*, Mar. 28, 2019, https://cathe.com/the-science-of-muscle-memory-and-how-it-works-in-your-favor/.

16. Ibid.

17. Charlie Wood, "What Is Gravity?" *Space*, Aug. 21, 2019, https://www.space.com/classical-gravity.html.

18. Adrian Budhram, "How Does Gravity Affect Basketball?" *SportsRec*, Jul 8, 2011, https://www.sportsrec.com/475170-gravity-affect-basketball.html.

19. Ibid.

20. Anne Marie Helmenstein, Ph.D., "What Is the Difference Between Mass and Volume?" *ThoughtCo.*, Oct. 7, 2019, thoughtco.com/difference-between-mass-and-volume-609334.

21. Ibid.

22. Brian Kross, "What Do We Use to Measure Mass?" *JLab Science Eduction*, https://education.jlab.org/qa/mass_01.html.

Chapter 2:

1. Lynn Aziz, Yazan Dirani, and Dana El Khatib, "Basketball Scoreboard History, Development, and How It Helps," *Academia.* Nov. 25, 2014, https://www.academia.edu/12674032/Basketball_Scoreboard_History_development_and_how_it_helps.

2. Ibid.

3. "Nissen Scoremaster 'Cadet' Scoreboard." *Smithsonian*, https://americanhistory.si.edu/collections/search/object/nmah_1372122.

4. https://basketballstem.weebly.com/basketball-technology
.html.
5. "Nissen Scoremaster 'Cadet' Scoreboard."
6. Ibid.
7. Lynn Aziz, Yazan Dirani, and Dana El Khatib, "Basketball Scoreboard History, Development, and How It Helps."
8. Ibid.
9. Tom Harris, Chris Pollette, and Wesley Fenlon, "How Light Emitting Diodes (LEDs) Work," *Howstuffworks*, https://electronics.howstuffworks.com/led.htm.
10. Noah Trister, "How Cutting-Edge Technology Helps Basketball Players Shoot." Associated Press, Mar. 20, 2019, https://apnews.com/ebe46bb1a07e4957a01629 12bbdfd997.
11. Juliann Allen, "Six Pieces of Wearable Sports Technology Every Basketball Player Should Know," *Teams of Tomorrow*, Aug. 8, 2016, https://teamsoftomorrow.com/six-pieces -wearable-sports-technology-every-basketball-player -know/.
12. "DribbleUp Smart Basketball and Smart Soccer Ball Deliver Effective Drills Using Augmented Reality," *Fitness Gaming*, Jun. 28, 2018, https://www.fitness-gaming.com /news/home-fitness/dribbleup-smart-basketball-and -smart-soccer-ball-deliver-effective-drills-using -augmented-reality.html.
13. Ted Kritsonis, "DribbleUp Review," *Digital Trends*, Sep. 19, 2016, https://www.digitaltrends.com/gadget-reviews /dribbleup-review/.
14. "What's the Difference Between AR, VR, and MR?" *Franklin Institute*, https://www.fi.edu/difference-between -ar-vr-and-mr.
15. Russel Karp, "How Technology is Changing the Sports Fan Experience," *Medium*, Apr. 5, 2019, https://medium.

com/swlh/how-technology-is-changing-the-sports-fan-experience-6f32a5bf921d.

16. ABC News, "NBA Teams Enhancing Fan Experience with High-Tech Replays," *Communications of the ACM*, Nov. 18, 2019, https://cacm.acm.org/news/240980-nba-teams-enhancing-fan-experience-with-high-tech-replays/fulltext.

17. Russel Karp, "How Technology is Changing the Sports Fan Experience."

18. Preston Gralla, "Technology Takes Center Court in the Basketball World," *Hewlett Packard Enterprise*, Mar. 4, 2019, https://www.hpe.com/us/en/insights/articles/technology-takes-center-court-in-the-basketball-world-1903.html.

19. Matt McLaughlin, "Technology Is a Game Changer for the NBA." *BizTech Magazine*, Oct. 16, 2018, https://biztechmagazine.com/article/2018/10/technology-game-changer-nba.

20. Ibid.

21. Preston Gralla, "Technology Takes Center Court in the Basketball World."

22. Mark Wilson, "Moneyball 2.0: How Missile Tracking Cameras are Remaking The NBA," *Fast Company*, Jun. 20, 2012, https://www.fastcompany.com/1670059/moneyball-20-how-missile-tracking-cameras-are-remaking-the-nba.

23. Marshall Zweig, "Technological Innovations That Changed the NBA Forever," *Bleacher Report*, Mar. 16, 2013, https://bleacherreport.com/articles/1568161-technological-innovations-that-changed-the-nba-forever.

24. Ibid.

25. Ibid.

26. Associated Press, "Basketball Refs Using Clock Technology to Get It Right," *USA Today*, Apr. 3, 2019, https://www.usa

today.com/story/sports/ncaab/2019/04/03/basketball
-refs-using-clock-technology-to-get-it-right/39294455/.

27. Ibid.

Chapter 3:

1. "Spalding History," *Spalding Basketball*, https://www
.spalding-basketball.com/de/about-spalding/history/.

2. Kirschner, Chanie. "Here's How Basketballs, Baseballs
and Footballs Are Made." *MNN Lifestyle—Arts & Culture*,
Nov. 7, 2017, https://www.mnn.com/lifestyle/arts-culture
/stories/how-basketballs-baseballs-footballs-made.

3. Sam Fortier, "9 Things to Know About the Life of an NBA
Basketball," *Esquire*, Jun. 9, 2015, https://www.esquire
.com/sports/news/a35587/nba-basketball-origin/.

4. "Behind Every Bucket: How the Official NBA Basketball
Becomes Game Ready," *Spalding*, Apr. 17, 2017, https:
//www.youtube.com/watch?v=zi-7Dw4BMqU.

5. Margaret Hoffman, "What's the Purpose of the Dots on a
Basketball?" *Mental Floss*, Apr. 1, 2015, https://www.mental
floss.com/article/62591/whats-purpose-dots-basketball.

6. "Why Are Basketballs Orange?" *Children's Museum
of Indianapolis*, Nov. 10, 2014, https://www.childrens
museum.org/blog/why-are-basketballs-orange.

7. Paul Steinbach, "Manufacturers Continue to Improve
the Basketball Goal," *Athletic Business*, Mar. 2005, https:
//www.athleticbusiness.com/Gym-Fieldhouse/goal
-oriented.html.

8. John Keilman, "This Gadget Really Was a Slam-Dunk,"
Chicago Tribune, Apr. 4, 2005, https://www.chicagotri
bune.com/news/ct-xpm-2005-04-04-0504040109-story
.html.

9. Paul Steinbach, "Manufacturers Continue to Improve the
Basketball Goal."

10. Joseph Eitel, "The Basketball Hoop: A History," *SportsRec*, Nov. 16. 2018, https://www.sportsrec.com/6542805/the -basketball-hoop-a-history.

11. Wyman Fraley, "Evolution of the Basketball Hoop." *Timetoast*, https://www.timetoast.com/timelines/evolution -of-the-basketball-hoop.

12. Robert G. Logan, William George Mokray, and Larry W. Donald, "Basketball." *Encyclopaedia Britannica*, Mar. 20, *2020*, https://www.britannica.com/sports/basketball.

13. Bryan Maddock, "Basketball Rims & Nets." *Dimensions Guide*, Sep. 22, 2109, https://www.dimensions.guide /element/basketball-rims-nets.

14. Charles W. Bryant, "What Makes Sports Flooring Different?" *How Stuff Works*, https://home.howstuffworks.com/home -improvement/home-diy/flooring/what-makes-sports -flooring-different1.htm.

15. Tim Newcomb, "Facts about floors: Detailing the Process behind NBA Hardwood Courts," *Sports Illustrated*, Dec. 2, 2015, https://www.si.com/nba/2015/12/02/nba-hardwood -floors-basketball-court-celtics-nets-magic-nuggets -hornets.

16. Charles W. Bryant, "What Makes Sports Flooring Different?"

17. Brendan Bowers, "From Chuck Taylor to LeBron X: Year-by-Year Evolution of NBA Sneakers," *Bleacher Report*, Feb. 7, 2013, https://bleacherreport.com/articles /1519230-from-chuck-taylor-to-lebron-x-year-by-year- evolution-of-nba-sneakers.

18. Kurt Badenhausen, "The NBA's Richest Shoe Deals: LeBron, Kobe and Durant Are Still No Match For Michael Jordan." *Forbes*, Aug. 28, 2019, https://www.forbes.com /sites/kurtbadenhausen/2019/08/28/the-nbas-richest

-shoe-deals-lebron-kobe-and-durant-are-still-no-match
-for-michael-jordan/#576a9fa13d02.

Chapter 4:

1. Austin, Michael. "Building the Perfect Arc in Your Shot." *Winning Hoops*, https://winninghoops.com/article/building -the-perfect-arc/.
2. "Is a Higher Arc Really Better?" *Noah Basketball*, Dec. 3, 2017, https://www.noahbasketball.com/blog /is-a-higher-arc-really-better.
3. "Basketball Court Dimensions," *Sports Know How*, https: //sportsknowhow.com/basketball/dimensions/basket ball-court-dimensions-diagram.html.
4. "Everything You Need to Know About Basketball Court Dimensions," *PROformance Hoops*, https://proformance-hoops.com/basketball-court-dimensions/.
5. "Guide to Basketball," *BBC*, http://news.bbc.co.uk/sport2 /hi/other_sports/basketball/4184748.stm.
6. Dan Kopf, "Data Analytics Have Made the NBA Unrecognizable." *Quartz*, Oct. 18, 2017, https://qz.com /1104922/data-analytics-have-revolutionized-the-nba/.
7. Jeff Haefner, "9 Stats That Every Serious Basketball Coach Should Track," *Breaktrhough Basketball*, https: //www.breakthroughbasketball.com/stats/9_stats _basketball_coach_should_track.html.
8. Keely Flanagan, "Basketball's Shot Clock: A Brief History." *WBUR*, Apr. 22, 2015, https://www.wbur.org/only agame/2015/04/22/nba-shot-clock-history-basketball.
9. Ethan Trex, "Why Is the NBA Shot Clock 24 Seconds?" *Mental Floss*, May 21, 2015, https://www.mentalfloss .com/article/13061/why-nba-shot-clock-24-seconds.
10. Jason Dachman, "NBA Tipoff 2019: Turner Sports Debuts New On-Court Virtual Shot Clock for *NBA on TNT*," *Sports*

Video Group, Oct. 23, 2019, https://www.sportsvideo
.org/2019/10/23/nba-tipoff-2019-turner-sports-debuts
-new-on-court-virtual-shot-clock-for-nba-on-tnt/.

11. Tom Huddleston Jr., "These Are the Highest Paid Players in the NBA Right Now," *CNBC*, Oc. 22, 2019, https://www .cnbc.com/2019/10/22/highest-paid-players-in-the-nba-right-now.html.

12. "NBA Player Salaries—2019-2020," *ESPN*, http://www.espn .com/nba/salaries.

13. Steve Fiorillo, "What Is the Average Income in the U.S.?" *The Street*, Feb. 11, 2020, https://www.thestreet.com /personal-finance/average-income-in-us-14852178.

14. "White Coats Can Jump: How Science and Technology are Shaping the Future of Basketball," *The Solidworks Blog*, May 14, 2019, https://blogs.solidworks.com/solid worksblog/2019/05/science-technology-future-basket ball.html.

15. Tom Ziller, "Why Players Make So Much Dough, As Explained by the NBA Money Cycle." *SBNation*, Nov. 18, 2011, https://www.sbnation.com/nba/2011/11 /18/2570807/nba-lockout-money-cycle-hook.

16. "Air Jordan 1/Designer:Peter Morre/Released: 1985 /Original Price: $65." *Foot Locker*, https://www.footlocker .com/history-of-air-jordan.html.

17. Michael D. Sykes II, "How Do NBA Shoe Deals Work?" *SBNation*, Jun. 22, 2017, https://www.sbnation .com/2017/6/22/15843134/nba-shoe-deals-nike-adidas -under-armour-guide.